恐龙化石

Dinosaur Fossils

[英] 露丝·欧文/著

刘颖/译

汉英对照
恐龙科普

江苏凤凰美术出版社

全家阅读
小贴士

★ 每天空出大约10分钟来阅读。

★ 找个安静的地方坐下，集中注意力。关掉电视、音乐和手机。

★ 鼓励孩子们自己拿书和翻页。

★ 开始阅读前，先一起看看书里的图画，说说你们看到了什么。

★ 如果遇到不认识的单词，先问问孩子们首字母如何发音，再带着他们读完整句话。

★ 很多时候，通过首字母发音并听完整句话，孩子们就能猜出单词的意思。书里的图画也能起到提示的作用。

最重要的是，感受一起阅读的乐趣吧！

扫码听本书英文

Tips for Reading Together

• Set aside about 10 minutes each day for reading.

• Find a quiet place to sit with no distractions. Turn off the TV, music and screens.

• Encourage the child to hold the book and turn the pages.

• Before reading begins, look at the pictures together and talk about what you see.

• If the child gets stuck on a word, ask them what sound the first letter makes. Then, you read to the end of the sentence.

• Often by knowing the first sound and hearing the rest of the sentence, the child will be able to figure out the unknown word. Looking at the pictures can help, too.

Above all enjoy the time together and make reading fun!

Contents 目录

史前骨骼
Prehistoric Bones

一天，两名科学家在岩漠里行走。

他们发现一些大骨头从岩石里冒出来。

One day, two **scientists** were walking in a rocky **desert**.

They saw some big bones sticking out of a rock.

骨头 bones

那是霸王龙的骨头！

They were the bones of a Tyrannosaurus rex!

霸王龙
Tyrannosaurus rex
(tie-RAN-oh-SAW-rus rex)

这些史前骨骼大约有6600万年的历史。

The **prehistoric** bones were about 66 million years old.

骨骼化石
Fossil Bones

恐龙的骨骼被称为化石。
骨骼化石又黑又硬，
因为它们已经变成了岩石！

The bones of a dinosaur are called
fossils.
Fossil bones are dark and hard
because they have turned to rock!

新鲜的骨头
fresh bones

霸王龙足部化石
fossil T. rex foot

恐龙的牙齿也变成了化石。霸王龙的牙齿是所有恐龙里最大的。

A dinosaur's teeth also became fossils. T. rex had the biggest teeth of any dinosaur.

霸王龙牙齿
T. rex teeth

石化
Turning to Rock

恐龙骨骼的石化需要几千万年。

It took tens of millions of years for a dinosaur's bones to become rock.

起初，一头霸王龙死在了河边。

First, the T. rex died near a river.

然后，其他动物吃掉了霸王龙尸体上的肉。
Then other animals ate the meaty parts of the T. rex's body.

这头霸王龙也许是老死或病死的。
The T. rex may have been old or sick.

冲刷
Washed Away

没过多久，这头霸王龙的尸体只剩下一副骨架。

Soon, all that was left of the T. rex was its bones.

接着，雨水将骨头冲进了河里。
骨头陷入河底的淤泥。

Then rain washed the bones into the river.
The bones sank into the mud at the bottom of
the river.

11

岩石恐龙
A Rock Dinosaur

数千年过去了。

Thousands of years passed by.

霸王龙骨头上方的淤泥越来越多。

The T. rex became buried under more and more mud.

后来，河流干涸，淤泥变成了岩石。

Then the river dried up and the mud turned to rock.

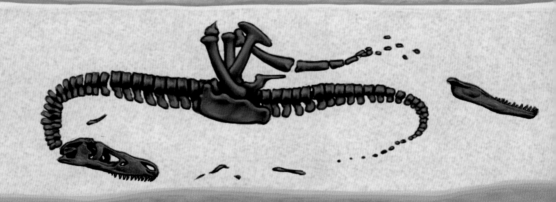

霸王龙的骨头也变成了岩石。

The T. rex's bones turned to rock, too.

重见天日
The Fossils Appear

几百万年过去了。

霸王龙的骨头被越来越多的岩石包裹住。

又过了几百万年，风霜雨雪使岩石破裂。

一天，霸王龙的骨头从岩石里冒了出来！

Millions of years passed by.

More rock covered the T. rex's bones.

After millions more years,

wind, rain and snow made the rock break up.

One day, the T. rex's bones stuck out from the rock!

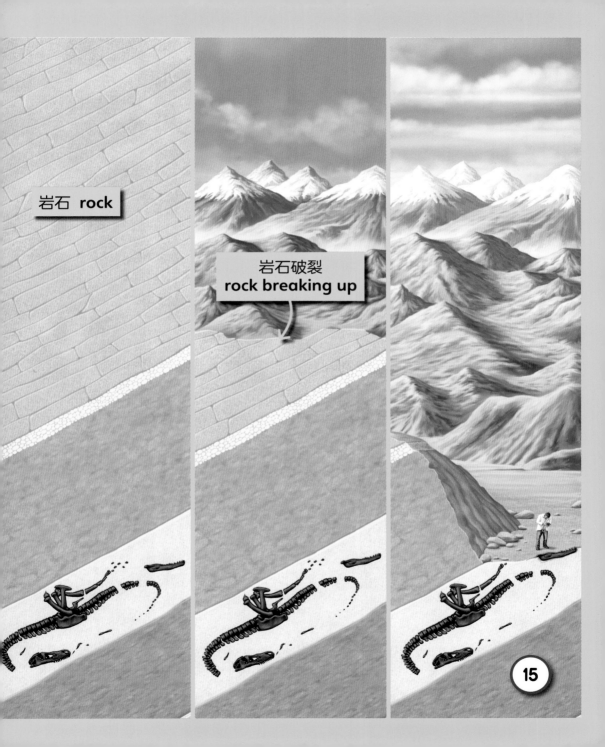

史前拼图
A Prehistoric Jigsaw

当科学家发现岩石里的化石时，
他们将它挖了出来。

When scientists see a fossil
in the rock, they dig it out.

霸王龙的皮肤
T. rex skin

科学家还发现了恐龙皮肤的化石。
Scientists have also found fossils of dinosaur skin.

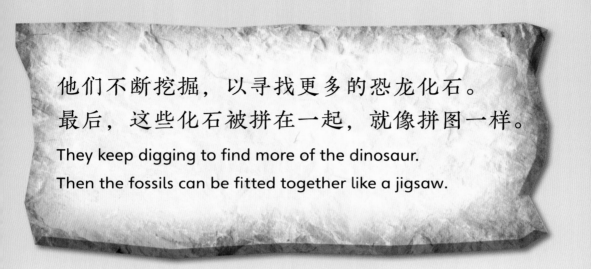

他们不断挖掘，以寻找更多的恐龙化石。

最后，这些化石被拼在一起，就像拼图一样。

They keep digging to find more of the dinosaur.

Then the fossils can be fitted together like a jigsaw.

史前粪便
Prehistoric Poo

科学家不只是发现了恐龙的骨骼和皮肤。他们还发现了恐龙的粪便。粪便化石被称为粪化石。

Scientists don't just find dinosaur bones and skin.

They also find dinosaur poo.

Fossil poo is called **coprolite**.

粪化石
coprolite

通过观察粪化石里的植物或骨头，科学家就能知道恐龙吃了什么。

Scientists can see bits of plants or bones inside the coprolite. This tells them what dinosaurs ate.

岩石脚印
Rocky Footprints

科学家还发现了恐龙的脚印。

Scientists also find dinosaur footprints.

食肉性恐龙的脚印
a meat-eating dinosaur's footprint

脚印化石是恐龙在淤泥里行走时留下的。

A fossil footprint is made when a dinosaur stepped in mud.

后来，淤泥变成了岩石。
Then the mud turned to rock.

食草性恐龙的脚印
a plant-eating dinosaur's footprint

这个脚印长106厘米。
它属于一种巨大的食草性恐龙。
This footprint is 106 cm long.
It belonged to a huge plant-eating dinosaur.

词汇表 Glossary

粪化石　coprolite

动物粪便历经很长时间
变成的化石。
The poo of an animal from long
ago that has become a fossil.

荒漠　desert

干燥的陆地，植物
罕见且降雨稀少。
Dry land with few plants
where very little rain
falls.

化石　fossil

存留在岩石中几百万年前
的动物和植物的遗体。
The rocky remains of an animal or
plant that lived millions
of years ago.

史前　prehistoric

人类开始记录历史前的一段时间。
A time before people began recording history.

科学家　scientist

研究自然和世界的人。
A person who studies nature
and the world.

23

恐龙小测验 Dinosaur Quiz

① 霸王龙生活在多久以前？
How long ago did Tyrannosaurus rex live?

② 恐龙的骨骼叫什么？
What are the bones of a dinosaur called?

③ 哪种恐龙的牙齿最大？
Which dinosaur had the biggest teeth of any dinosaur?

④ 为什么科学家只发现了恐龙的骨骼而不是身体？
Why do scientists find dinosaur bones but not dinosaur bodies?

⑤ 你愿意捡起一块粪化石吗？
Would you pick up a lump of coprolite?